In this four-Dimensional World we call, "Reality" there are those of us who hardly ever think outside the box. Yet there are other World's that exist in other Dimensions that are just beyond our common sensory perceptions.

We exist daily within the four physical spatial dimensions of up, down, left, and right, along with time…

However, many of us may not know or even consider that concurrently there are parallel planes or extra Dimensions inhabited by others.

Whether one chooses to believe or not, my books contain compelling evidence that there is more to reality than meets the eye.

Now come with me on a short journey around Cape Ann, soon you will believe, as I do, that,

They're Here and Watching!

About the Author

The books I am writing are autobiographies of my life. I have experienced Supernatural occurrences since I was a young girl. I now share my stories with you. I believe my heritage holds the key to my awareness. My father's mother, Rosebud Burke, was half Cree Indian. She had a light caramel complexion, with long, beautiful, curly jet-black hair. Her husband, my grandfather, whom I never met, was Haitian; this combination alone makes for strong spiritual bonds.

*We have all heard stories of Indians, and Haitians spiritual connections. Most of us will admit that Spirits or Ghost do exist, (some choose not to believe what they cannot explain). I, however, believe people are missing what these Spirits or Ghosts are trying to tell us. I believe the message is clear. That message is "**The Spirit World does exist, so it stands to reason that God and his son Jesus Christ do exist! It's time we realized that we need to love and respect one another, no matter***

what the color of our skin or our status in life. That is how we treat ourselves and others really does matter. That we are all being watched and judged, and how we live here determines our faith in the Afterlife. *My experiences along with my pictures will show you that we are hardly ever alone. I've tried to dismiss a lot of the things I've seen and heard throughout my life, but here in New England the Spirits are so active and so unique, there just is no denying, that, They're Here and Watching!*

Antrella Grice

They're Here, And Watching II

"The Peeper's"

Among these ancient headstones in this cemetery,

there is a stairway with an unknown destination.

Join me for a brief tour of Cape Ann where every

stone holds a story. As spirits are present in various

places throughout the Island.

I serve as a representative for the spirits that I have encountered. The information in their Dimension could be important for our future. Leave your doubts behind, open up your mind and walk with me.

Contents

House **13**

The Peeper's

My Thoughts…

Fifth Revision – July 2025

They're Here, And Watching II

"The Peeper's"

By Antrella Grice

ISBN-13: 978-1475273649

ISBN-10: 1475273649

These photos were taken seconds apart; note the change in light.

This is located along the second path to Flat Rocks.

Introduction

Are ghosts, spirits, and orbs genuine phenomena? Are they real? Or could it be just our overactive imaginations? We ask ourselves, "What are the shadows moving about the room?" and that gust of wind that rushes past us. How about the sudden scent of a loved one's perfume or cologne that comes from out of nowhere? These supernatural events suggest an Afterlife, of this, I am sure. On Cape Ann, I have connected with Spirits from other Dimensions who are no longer alive to us but still exist. The boundaries between worlds or dimensions are thinning, causing overlaps in space and time. Some entities can observe us and attach themselves to our living world, triggering unusual events like doors opening by themselves.

Objects disappearing unexpectedly, only to reappear moments later.. Some have harmful intentions. Therefore, keep faith and remember,

They're Here and Watching.

In dedicating this book, I acknowledge the spirits who have revealed themselves to me throughout the years. Without these experiences and pictures, documenting ghostly encounters would not be as effective. The most unique picture is presented last, aiming to convey what words may not express fully. Enjoy my experiences and enjoy the pictures, they will no doubt blow your mind and make you wonder.

Antrella Grice

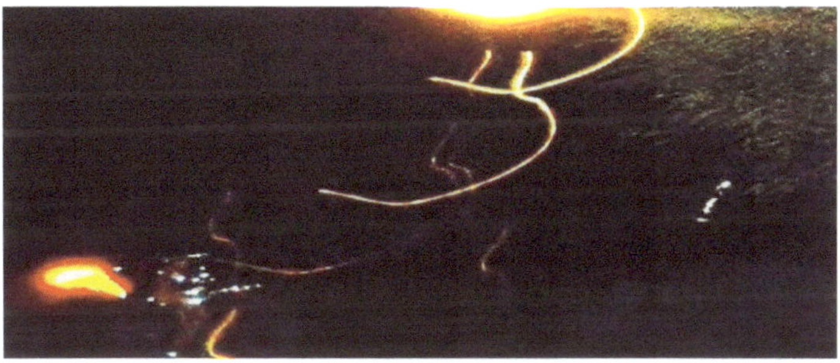

Strange Energy Lights in Rockport, MA summer evening 2011

One

1991 - Chicago, Illinois

At seven months pregnant, I struggled with insomnia, and this Saturday night was no exception—I just couldn't get comfortable in bed. Herman, my son's father, moved the futon from the living room and laid it out on the floor. I anticipated that this would finally allow me to rest.

We lived in Chicago near Lake Shore Drive, within Belmont Avenue on the north side. Belmont Avenue is lively with weekend partygoers. Between Lakeshore Drive and Lincoln Avenue, Belmont offers plenty of entertainment options. There are several movie theatres, fine dining options, and unique shops catering to various tastes. Our blues clubs are captivating, and we also have some hidden dance clubs. Additionally, there are some Hard and Soft Rock spots, but Lake Michigan is notable for its natural features. The calm blue waters provided comfort. During my adolescence, I frequently spent summer days at Belmont Rocks. I would descend the rocky terrain to find a secluded spot and watch the waves while reading my novels. Although I was born on the west side of Chicago, my family moved to the north side when I was eight, where I lived until moving to New England in 2005.

It was a warm August night and Belmont was busy as usual, with an indistinct atmosphere. I lay there hearing groups of people laughing as they walked by, their spirits affected by drinks. never realizing the real Spirits there with them. Impatient horns blast as anxious people travel, sounds of taxi whistles and sporadic altercations accompanied by police sirens permeated the atmosphere. I could envision the red fire trucks as they went screaming past. Rushing to someone's rescue. The sounds of a city. Among all that noise below, I felt secure.

Lying on my side and looking into the night, I remembered being part of that crowd months ago. My friends and I established a routine of visiting various restaurants on Fridays and going to clubs on Saturday nights. That summer, we visited The China Club, a nightclub located downtown in an old Victorian building with three levels. In the 90s, it was a popular place for dancing and socializing. A large bar was in the center of the entrance, with two smaller bars in the corners to reduce wait times. The first floor housed a jazz club with nightly buffets for the after-work crowd. There were two clubs on the second floor and a large room with live bands on the third floor for dancing. If you want a break from dancing, the lower level offers pool tables, video games, and three fast-food restaurants for sitting and chatting.

Sweet memories filled my thoughts, but the best was yet to come. I smiled, rubbing my belly. The birth of my first child was only a few months away, and I looked forward to meeting them. I decided it was time to rest, understanding that I would not sleep if I continued to look out the window. I lay on my right side, used my arm as a pillow, and closed my eyes to sleep. Belmont and Lakeshore Drive feature beautiful vintage high-rise buildings. Our five-story building features extended hallways with dim lighting. The carpet was worn, blood red and black, covered with old footprints. The elevator had a sliding metal gate, common in the twenties, which made a sound throughout the building whenever it closed. The building's interior design reflects what was once a luxurious hotel. The cage on the right side of the lobby, where guests once registered, now stands locked, filled with cobwebs and memories. My footsteps echoed in the lobby. As I walked to the elevator, it felt like the building was still operating as it did in the past. I sometimes felt watched by invisible eyes in the cage at night. At times, I felt it necessary to sign in, and the cage and elevator caused some concern. I recalled the tales my grandparents and parents told me as a child. Although I was aware of the existence of the Spirit World, I was hesitant to fully embrace it and attempted to suppress those thoughts. It is preferable to overlook and ignore what I understand to be true. Saturday night will be life-changing for me. The concept of an afterlife would hold true for me, leading to a profound shift in how I perceive life and our existence. I gained a new perspective, as it were.

Leaving behind Belmont's traffic and noise, I tried to sleep but suddenly felt a strange sensation. I had an awareness of another presence. I sensed an unidentified presence, and my initial reaction was to protect my abdomen. I felt a sudden chill. While lying down and rubbing my abdomen, I noticed a woman in the kitchen doorway dressed in a white flowing gown, with long, voluminous, wavy hair that nearly reached her waist. I was positioned approximately five feet from her location. Her dress appeared to be an elegant ballroom gown from the 19th century, suggesting that it was once quite beautiful. Featuring a full skirt and jeweled bodice.

She remained in the doorway, hovering and surveying the living room. Her appearance is still clear in my memory today. She glided towards me while I lay frozen; she hadn't noticed I was watching. She chose to navigate around me rather than float above or pass directly through me. As she moved past my extended arm, I sensed a chill; it seemed she was reluctant to cross over my awaiting abdomen. A small animal followed her. The animal was a dog. Its small size and shaggy hair resembled a terrier. Initially, I was able to see through her and her dog. However, as they lingered at the doorway and proceeded further into the room, their appearance became increasingly solid. The situation was so startling that I remained immobile until I gathered the courage to gradually turn my head, being careful not to produce any abrupt sounds.

With my arm extended, I turned my head to observe her as she bent over an end table, which was brought from Rockport, Massachusetts, along with the crystal figurines she was admiring. The ballerina captured her attention, coincidentally being my preferred choice as well. It appeared she might have been the former owner of the item, many years ago. She remained poised above the ground as she leaned forward, gently stroking the crystals. She was smiling, which suggested that it reminded her of a more pleasant time.

While I was observing her dog sniff around the room, I had not noticed that she had turned her head and was looking at me. I recall how she appeared suspended in midair, without any movement. The eerie stillness made my skin crawl. When our eyes met, her expression turned to anger.

Her lips parted, revealing decayed teeth. She sneered and hissed, sending a foul breath my way. In a rapid motion, she moved through me with such speed that I felt her cold and anguished presence permeate my entire body. I was concerned about my unborn child, considering the possibility that he could sense her cold energy. As the ghostly pair entered the kitchen, pots and pans on the radiator sounded as if they were tossed to the floor. I was so surprised and taken aback that I found myself unable to move. Concerned about the possibility of premature labor or miscarriage, I regulated my breathing and remained motionless until my heart rate returned to normal. After a long wait, I turned my head, stood up slowly, looked towards the kitchen, and decided not to investigate. Instead, I entered the bedroom where Christopher's father was sleeping.

My initial thought was to wake him and inform him of the recent events. However, anticipating disbelief, I proceeded to the bathroom. I turned on the light, closed the door, washed my face with cold water, and looked at myself in the mirror. I shook my head in complete disbelief, unsure of what to conclude. However, I was certain that I was not dreaming and had indeed encountered an actual spirit or ghost, although I was uncertain which. I initially regarded ghosts as malevolent and intimidating, while spirits were benevolent and supportive. This apparition exhibited a benign demeanor initially, but subsequently it did not maintain this gentle nature. I am unsure how long I spent in the bathroom, but I eventually felt very tired and needed to lie down. I gradually opened the bathroom door. Observing Christopher's father resting peacefully, I felt inclined to lie down beside him for a sense of security. Although I was reluctant to return to the living room, the discomfort of the waterbed made it impossible for me to sleep. I switched off the bathroom light and cautiously entered the living room, searching for any indication of the ghostly presence. It appeared they had already departed for the night. I lay down once more, closing my eyes tightly. I recited the Lord's Prayer repeatedly until I eventually fell asleep.

The next morning, I was awakened by the scent of bacon and coffee. Herman was in the kitchen preparing breakfast. I sat up and started describing the ghostly pair to him. As I reached the section describing the clatter of pots and pans hitting the floor, I halted abruptly. I was stunned to see pots and pans sitting on the radiator in the kitchen. My legs felt weak, and I leaned against the wall for support. The pots and pans confirmed I had seen a real spirit last night. She evidently passed through the radiator and agitated the pots and pans in a display of anger. Herman expressed keen interest in this spectral pair and inquired about the appearance of the woman. I described her long, flowing hair to him. I also mentioned the exquisite gown adorned with jewels of various colors that reflected brilliantly off the walls. Herman's eyes widened with amazement as he entered the living room and picked up the crystal ballerina. His reaction to my story was unexpected. I expected him to dismiss my concern, but he turned toward me and said, "I wish you had woken me last night; I often feel like someone is here with me, watching me. "This observation marked the beginning of my understanding of the phenomena that exist around us daily. Since then, I have experienced heightened senses, and my spiritual awareness has been awakened. I have observed and documented various instances of spirits or ghosts through visual, auditory, and physical encounters. I now present these findings to you.

Strange Energy Lights in Rockport, Massachusetts.

Two

1994 - Old Farm Inn- Rockport, Massachusetts

My son Christopher and I have consecutive birthdays: he was born on October 3, 1991, and I on October 4. Additionally, I share my birthdate with my sister Phyllis, who was born two years earlier. The three of us have a unique bond and often sense when one of us is facing difficulties, prompting calls or texts. Christopher and I share a strong bond. His father is from Rockport, Massachusetts.

I met Herman at Kingston Mines, a Blues club in Chicago. He was a door attendant with a New England accent and blue eyes. I knew he would help me discover another part of the world that I would eventually call home. After dating a few months, he invited me to spend Christmas with his family in Rockport *(my bags were packed right after our first date).* We drove to Massachusetts, and this being my first road trip and my first time traveling to the east coast, the ride seemed like it took forever. During the day, the scenery was beautiful. When night fell, I did not like the dark winding roads, hills, and mountains. When you grew up in a big city like Chicago, dark roads and dark streets meant trouble and my guard went up. However, when we finally reached exit 11 into Gloucester, all my fears subsided. My eyes had never seen a town more beautiful and festive. We stopped at a real New England pub, and it felt like I was in a dream, as though I was floating, and every eye was on me. After we warmed our bones, we drove into Rockport and my heart fell to my feet. I knew I wanted to live here. The three of us were now taking our first trip to Rockport for a visit during the fall of `94. We stayed at a bed and breakfast restaurant, called Old Farm Inn, a converted farmhouse. Herman was friends with the owner who fixed us up with a great room. It was on the top floor of one of the buildings, which had been part of the original barn. It had the typical small television set, a desk and a chair or two, and of course a bathroom. The kitchenette was useful with a 3-year-old in tow. It felt like a home away from home. On our first night, we had dinner in the main house dining room. Afterwards we drank hot apple cider and sat in the common area taking pictures and talking with some of the other guests. My only previous experience of New England was in movies and reading books that took place here. Being here was wonderland overload for me.

The next morning after showers, the three of us headed down to breakfast. The owner of the Inn had the most pleasant, warm and friendly smile I had ever seen. He came over and welcomed me to Rockport, and to his Inn. He asked how we found our room and to let him know if there was anything he could do to make us more comfortable. After chatting for a while, we thanked him and were off to visit relatives. Later that evening we returned to the Inn with lobsters given to us for a lobster fest. This would be my first time eating real New England fresh lobster and I was like a kid in a candy store. I managed to tear through two large lobsters and a few extra claws that night. We returned to our room around 10:00 pm that evening, and Christopher was already asleep, and Herman carried him back to the room. Putting our son down to bed, we went out on the small balcony where we took a few pictures. We talked for a short while but wanted to get some rest since breakfast is very early at the Inn. Staring out the small window into the darkness, (*Chicago has many streetlights*) it reminded me of a scene from a movie I had recently watched. The scene was set deep in the woods of New England. A house sat in the middle of the woods; the nearest neighbor was half a mile away. The camera panned to a second-floor window, where curtains are blowing in and out. A woman sat at her vanity brushing her hair. She was crying. She heard a howl outside in the distance. She paused and listened, silence. She began to brush her hair, and then she heard another howl, this time a bit closer. She still does not do anything, except glance at her reflection in the mirror. Just then she realizes there is something right outside her window, as she stands and begins to walk toward the window to close it, it's too late, the wolf jumps through… and well, you know the rest. I tried to stop thinking long enough to get to sleep when suddenly there were muffled voices, and what seemed like an endless number of people walking around upstairs. I smiled at myself and slowly got out of bed, being careful not to wake my two boys. I tiptoed into the bathroom and checked myself. I got into my jeans and t-shirt, slipped on my shoes, fluffed up my hair. I was ready to crash into a party, Chicago style. I went out the front door and down the stairs and looked up, there was a small window above ours, but there were no other stairs to this floor. "These old farmhouses", I said aloud, and shook my head. There had to be a door around back somewhere, but it was too dark to see. I went back into the room and into the kitchenette feeling somewhat puzzled, I cocked my head to one side listening to the party upstairs. There was a back door in the kitchenette, but it did not open. I peered through its window, and saw a trap door in the ceiling, but I thought that was a bit too much effort. I finally gave up and went back to bed and listened to the people upstairs for some time. Finally, sleep came rushing in; I would wait until morning and find out how to get upstairs.

Morning arrived and I jumped out of bed when I heard the shower going Christopher was sleeping, Herman was in the shower. I opened the bathroom door and began telling him about the party upstairs last night, and how I could not find a way up! He asked me how I knew it was a party. Was there any music he asked? The strange thing is

I did not remember hearing any music. He told me to ask our friend about it when we got down to breakfast, I could not wait. As our friend came to the table, smiling friendly as usual, I began to tell him about the party going on upstairs last night, and that I looked for a way up, but could not find one. He casually said, "There's no upstairs, it's just an attic, it was only the "Ghosts", and he walked away. I watched him greet two ladies at the next table with my mouth hanging open, eyes wide. I was speechless (*and anyone who knows me knows speechless I am not*). Herman looked at me with a smirk, and said, "You asked" and continued eating. That evening and the remainder of our stay, I made sure I was so tired I would not be able to keep my eyes open after 9:00pm.

This was my second encounter with Spirits as an adult, and I was especially intrigued with the New Englanders who talk of Ghosts with such nonchalance. I needed to understand why. As a young girl I enjoyed reading novels, the bigger the better, mysteries were my favorites. I was and still am a fan of Agatha Christy. I moved on to V.C. Andrews, with Flower's in the Attic series. In my early twenties, I became interested in the supernatural world and found Ann Rice to be the cat's meow. When I was in my late twenties, there came my Steven King fascination, and I have been a fan ever since. His description of the towns, and the people, New England towns and New England people drew me in. I now know my attraction to stories and movies about New England, and meeting Christopher's dad was no coincidence.

Visiting New England had only been a dream in my mind, especially as I turned each page of a Steven King book. Now, years later, I have been here, and it has changed my life forever. Being fascinated and terrified all at the same time, my destiny was set.

The morning before we headed back to Chicago, I took off alone for a walk in the woods in Pigeon Cove. Being from a big city, I am not the type to go off walking in the woods, alone or with others, but an unseen force pulled me there. As I stepped onto the path, I began to sing loud, and louder and louder still, taking in all the beauty around me. Then like a ton of bricks hit me, I felt like I was on fire, thoughts and ideas entered my mind that I had never thought of before. I would write about the Ghost and Spirits who seem to be drawn to me here. I thought perhaps they are trying to give us a message, and they chose me to get people's attention that ***"The Afterlife does exist"!***

Energy Lights in Rockport notice the faces on the right.

Three

Annisquam Alive

Have you ever visited a place that once you walked inside the hairs on the back of your neck tingled? Your breath was stuck in your throat. Perhaps you got an uneasy feeling, sensing a sinister presence. Well, I experienced all three at the same time at a house in Annisquam (*Gloucester, MA*). As soon as I walked through the door, my body became warm; I felt beads of sweat on my brow. As I continued into the kitchen, I felt a devious presence and I knew this house was breathing. I refer to this house as "Annisquam Alive", because the previous owners will not leave, and they do not want anyone else to live there either. Shortly after I moved to Gloucester, I met a couple of very nice people who became two of my best friends. They were roommates in this one particular house, which had stood vacant for some time. One of the friends retold a story he heard the previous owners had some mental challenges. The wife had undergone a lobotomy at some point, and she constantly played a certain tune on her piano, repeatedly, and it was driving her husband literally insane. He had tried to commit suicide by hanging himself in the cellar, which he failed to accomplish on more than one occasion. No one knew how or when couple died, but their activities live on harassing anyone who rents the house. Another story I heard about the house happened to a couple of movers hired to remove the furniture from the house. It was just another typical job for the movers. After moving a few pieces from the house, they both began to feel strange. They felt as though someone was watching them! When they were moving the piano through the front door, it began playing that haunting tune. Totally freaking them out, they dropped the piano and ran off not finishing the job.

My two friends had moved into the house, despite the stories. I do not think they really believed them, but that would soon change. They had a cookout one summer day and by the time I arrived, it was just getting dark out. I had been at another party earlier and was a bit tired. After a few hellos and hugs I went inside to take a quick nap. There were people inside of the house as well, so I went into the small bedroom off the kitchen. I closed the door and lay down. No sooner than I did this, I could hear someone inside the room talking lightly. It was a small bedroom, so if anyone had been in there, I would surely have seen someone. I thought maybe it was coming from the hallway, perhaps someone was waiting for the bathroom, which is right across the hall. I sat up a bit and tilted my head to hear better. Silence. I lay back down. A few minutes later, I began having trouble breathing. The air in the room became very heavy; it felt like the air itself was smothering me. I needed some fresh air, so I got up and opened one of the windows that faced the water. A great breeze of night wind swooshed inside the room. I let out a sigh of relief. I got back in bed to take that nap, only this time I saw someone standing in the corner by the table. Moonlight lit the room, and I saw the outline of two people. The man wore a suit, which looked wrinkled and a little big for him. The woman stood very close to him, her hair dark and short. They moved toward the bed holding my gaze. Their heads were somewhat bowed looking down at me. Their expressions were cold and hard. They did not like my being in that room, and they wanted me to know it. I was caught off guard, and frankly so tired that I was more annoyed than scared. I shook it off, went outside, and rejoined the party.

Months later, my friends were having a few people over to watch the sunset, and I was the first to arrive. One roommate was still at work and the other had a meeting in town. I was in charge to greet any early birds and doing a little setting up. I did not think much about the ghostly couple in the bedroom, not until fifteen minutes of being in the house alone. I was in the sitting room, watching television when I began to hear noises from the kitchen. Before I could react, I heard shuffling sounds upstairs, as if someone were dragging things around. I then knew exactly whom I was dealing with. I would not acknowledge them I thought, and I continued watching television. They were not having it! The noise in the kitchen got louder, the dishes in the sink began to click and clank. It sounded as if someone were doing dishes in anger. That is when my heart began to beat a bit faster. I rose up a bit from the chair I was in and peeked in the kitchen. I knew in my heart that I would not see anyone there. Not seeing anyone after hearing all that noise made me even more uneasy, I would rather have seen an intruder. I sat back and continued watching television, what could I do, I did not want to move. The shuffling and dragging upstairs continued. Someone was sweeping the floor in the kitchen now. Snapping my head from the kitchen, and then upward toward the ceiling was making me dizzy. Just as I was about to get up and leave, the noise stopped.

After eating dinner and watching the most amazing sunset over the Annisquam River, a few of us stayed over and decided to tell ghost stories. I was to sleep in the room off the kitchen. Relating to seeing the couple on the night of the barbeque, I was secretly hoping someone would change rooms with me. No takers. Then I told them what happened hours earlier while setting up for the party, with that said, it got the two roommates talking about some of their experiences. This house was very active, they said, and they were relieved someone else experienced the ghostly pair. Around two in the morning, we all headed off to bed. As tired as I was, I could not get to sleep. I kept tossing and turning until I finally got out of bed and went outside for a cigarette. The sky was so clear and was the bluest blue I had ever seen. Never was the sky this clear and blue in Chicago. I began to yawn uncontrollably. I put out the cigarette and went back inside to bed. I still could not get to sleep, and now I could see why, "*the pair*" was back. They stared at me with such hatred it was both haunting and frightening. That was it! I had enough. They could have their room! I got up, put on my shoes and headed for my car. The rest of the house in darkness, my friends were unaware of what was going on. The driver side of my car was facing the house. When I got in and started it up a light went on in the second-floor window. I could see the string swinging back and forth through the sheer curtain. This was strange because no one was on the second floor, and everyone in the house was asleep. I stared at the light in the window for a short time then I began backing out of the driveway. I shifted gears to turn onto the main road. I looked up at the house one last time before I pulled off, and just like that, the light went off. I was out done, "the pair" was glad that I was gone. I knew I would be back. The next day I called my friends and shared my dramatic exit. There was no denying the place was haunted; they said no one had gone upstairs that night. As summer approached, I made many friends in town. Gloucester was also one of the perfect vacation spots in New England. Almost everyone on the island has a boat and I spend most of my summer on the water.

Summer passed and the cold seasons were upon us, and I decided to drop in and see my two friends. Only one of my friends was home, the other was away on vacation. We put on some music and talked about our great summers. We talked late into the evening, so I decided to sleep over. I made myself at home after my friend went off to bed. I stayed up in the sitting room and watched a little television, falling asleep in the chair. I was awakened around 4:00 am by the smell of coffee. After talking with my friend a few minutes, I told him I would make sure to lock up after getting a few more winks. With a big hug, he was off to work. Again, I was there alone. I went back to the chair but thought about the soft bed; reluctantly I went to "the room" off the kitchen and lay down. With the glow of the moon shining through the window, I felt somewhat safe, so I closed my eyes to rest. In what seemed like minutes from lying down, my eyes sprang open! I could see the sun rising over the water and knew I had been asleep for hours. It was not the sun rising over the Annisquam River that had woken me. It was the sound of someone beating a pot with a spoon down by the water's edge. I could see in my mind's eye a woman in a dark dress with a flowery apron on; she was staring up at the house. I had to get up and look out the window to where I knew I would see her. However, before I could get up, she was right next to me clanging that pot in my ear! The noise was so loud I swear the walls were shaking. I thought I would lose my hearing. I jumped from the bed, screaming at her to "**LEAVE ME ALONE**"! Although, I thought for a split second, that it was her house, or she still felt it was! She had won. I left the house that morning and have only returned to take a few pictures for this book.

Seasons changed and I went to my favorite pub in Gloucester. Sitting at the bar was the roommate who was on vacation when I last visited the Annisquam house. I walked over and said hello, ordered myself a beer, and sat down next to him. When I turned and looked at him, he had a rather shaken expression in his eyes. He was the only one living in the house now. The other roommate had had enough of the ghostly pair and moved after his girlfriend had trouble sleeping there. When I asked him what was wrong, he told me a tale of sheer horror. I will never forget the look on his face as he told me what happened to him in the Annisquam house.

Since he was the only one living in the house, he took to sleeping in the bedroom off the kitchen. A few nights ago, he woke to the sound of heavy footsteps climbing the stairs to the second floor. He did not know what to do or what to think. He could hear the footsteps stomping downward now, getting closer to the bedroom where he was! Someone had broken in and was robbing him, was his thoughts. He heard sounds of things breaking, furniture being thrown about. He feared for his life and could not move. Then, without thinking, he climbed underneath the bed. The footsteps were getting closer; he could not be sure how many were out there, but they seemed to be in every room. The footsteps grew louder, and they echoed throughout the house, it sounded as if the house was being destroyed. Were they looking for him? He shook with fear. Then the house went silent. My friend was still underneath the bed frozen with fear. Could they be waiting for him to come out? After some time, he finally came from beneath the bed. He found something to use as a weapon, put his ear to the door and listened. He slowly turned the knob, the sweat running down his face, he was trying his best to keep as quiet as possible. The bedroom door gradually opened. He swallowed and emerged from the room. Looking around every corner as he walked, he saw no one, and nothing was disturbed. His mind went wild; he began to shake a bit more now. He walked slowly toward the kitchen door, he turned the knob and pulled it, locked from the inside. He was puzzled. He unlocked the door and stepped onto the enclosed porch; he walked over and tried the knob on the outer door. It was also locked. Now he really was panicking. The intruders had to be in the house still, perhaps in the cellar. He went back inside and made a rather panic-stricken search of the remainder of the house. Still nothing had been disturbed and there was no sign of anyone having been inside. This should have put his mind at ease, but he soon realized it was the "**ghostly pair".** He grabbed his keys and left the Annisquam house, never to sleep there again!"

The house sits vacant again but only vacant to the living.

Four

The Captain's House

One afternoon I was out with a few friends having lunch, and the conversation switched over to ghost. One of my friends, Melissa, started telling us, (*mainly me*) about the house she lived in. Captain Bartholomew, in the 1800s built the house and one night she had actually seen him. She told me the story, and it went something like this; (*One night she was awaken having a strange feeling come over her, I believe she said it was around 4:00 am. She had opened her eyes only to see her dog and cat at the foot of her bed, staring at the wall. The wall is all brick and has a nice sized fireplace; her television was to the left of the wall on a nightstand. She noticed how the animals were staring straight ahead as if they were in a trance. When she looked in the direction the animals were looking, there was someone walking past the fireplace. She had been sleeping soundly until now and had to blink a few times to be sure of what she was seeing. At first glance, she said he looked like a friend of ours, Dan. He was wearing the same cap and had the same beard, so she called out his name, and asked what he was doing there. However, she soon realized it was not Dan, but Captain Bartholomew himself. Although she should have been afraid, she felt no fear).* We decided to take a walk over to the house so I could take a few pictures; unfortunately, I only had my cell phone camera at the time. I did get a nice haze in front of the brick wall above the fireplace, which was a definite sign that something was there. A few years or so later Melissa moved into another house, not because of the visits from the captain, it was just time for a change. Helping her to clean up the place, was another friend of mine, Lisa who is very intuitive when it comes to Spirits. As Lisa was alone in one part of the house cleaning, someone slapped her on her bottom. When she turned around, there was no one there. It happened again before she left the house that day, and she knew it was the captain; although his presence was strong, Lisa felt no fear. A year or so after that I moved into the captain's house, in one of the first-floor apartments. It has two bedrooms, a smaller one off the kitchen, and the larger one, which has a nice deck attached toward the front of the house. All the doors in this house have latch key handles, and the kitchen and living room have exposed wood beams. I took the smaller bedroom off the kitchen, because it has more "*Old World*" character. The closet, although small, has exposed wood beams as well, with a skylight and ceiling fan, it felt very 19th century, and I felt right at home. With Christopher off to college, I was moving here alone so for two weeks after work, I moved whatever I could pack into my Ford Taurus. Finally, the move was finished, and I would spend my first night at the Captain's House. I walked into the house and immediately felt the captain's presence it was a little startling at first, but soon after taking my cats out of their carrying cases, I could feel the house relax. The captain liked me and was pleased at my genuine love and appreciation for his home. Only my quaint little New England home would become the "Amityville Horror" for a while. One Saturday morning I was in my room watching television, unpacking and hanging clothes in the small closet. While my back was turned, I heard a buzzing sound, as I turned around, I saw this huge black hairy fly inside of my closet. After jumping about 2 feet in the air and screaming, (***I really dislike flies***) I grabbed a magazine, and proceeded to chase this fly from room to room. After swatting the hairy, germy fly, I flushed him down the toilet, but before my hand left the handle, I heard another buzzing sound. I looked up and there, in front of me, was another fly with two of its friends. (*The words that came out of my mouth are not something I want to repeat, so use your imagination*). I ran from room to room looking for the flyswatter, cursing and knocking over boxes as I went. The hairy trio kept up with me, taunting me buzzing around me in unison. This meant war, and after about 15 minutes of out and out battle, I defeated the hairy crew. Now I could return to unpacking and cleaning up the mess I had made, I was feeling pretty good about myself. After about an hour or so later, I went to the bathroom, and to my surprise, I heard a buzzing sound again, "Dammit!" I screamed it was in the light fixture. I could not believe this! I finally called the owner hysterically complaining about the flies, wondering where they were coming from in my house. I only got a grunt from my property owner, not much, he could do…and I knew it was up to me to find and get rid of them. This went on for a few weeks; until finally I was rid of the last hairy beast. Life resumed.

I did my usual antiquing in and around Essex, and of course taking pictures, and enjoying The Captains house. Just as I had imagined, my taste for antique furniture, my "*Rococo*" style as some would say, was perfect in the Captain's House. (*My appreciation of antiques began as a little girl,*

something my father and I both shared.) Before I moved from Chicago, Illinois in 2005, one of my sisters bought a two-flat building in Alsip, Illinois. The previous owners left artwork, and my sister asked me if I was interested in any of the pieces before she got rid of them. One picture I liked very much which traveled to New England with me. The picture is of a little boy in his red velvet attire for the period along with his pets. His picture hung in my kitchen above the table next to my bedroom I could see him when I looked in that direction.

The next picture I believe is a Monk, I found him somewhere in Chicago, years before I got my little person. I hung him in the hall that connects the living room, kitchen and second bedroom so the two pictures were across the room facing one another.

Now prepare yourself for what you are about to see. There are two little legs under the table, orange in color; they are identical to the legs of the little boy. You can begin to see his hand holding the string. Notice the yellow glow. They are arms reaching out. The shape of the head resembles the shape of the Monk.

Take a closer look at those legs under the table and in the face of the Monk, his facial features begin to appear. Before that picture was taken of the glowing pair, I had a couple of friends over to the house, one being Lisa. It was a nice summer day, so we sat outside on the deck and enjoyed the fresh New England air. Afterwards I showed them my Pathe

Frères & Co., Phonograph and immediately they both got a weird feeling. We went into the kitchen and stood around talking, when something ran between the living room and the second bedroom. I saw it and so did Lisa (*who had been coughing ever since we came inside from the deck*). We gave each other that look of "did you see that?" and sort of shrugged it off. We went back into the living room, but the more we talked about the phonograph, the more Lisa coughed. Only now her eyes began to water, I saw that see she was suffering, she had to leave my house. Later the next day we talked about what happened and she told me she felt like someone was choking her. As to what ran between the rooms, she said it looked like the Monk. Many people do not like these two pictures, and I have had to take them down once I saw their glowing figures in the kitchen. What does this mean? Are certain types of pictures able to materialize? Alternatively, is it just my two pictures that are haunted were these two in fact real people and not just made up in someone's mind? Does this mean that once we depart from this World, and have had our picture painted we will be able to step off the frames as well? Several nights before I saw my pictures come alive, I knew the captain, or someone was in the closet in the living room. One evening while hanging up my jacket, I really felt a strong presence; I put the latch on and went about my business. A few hours later, I went to the bathroom and when I came out on the living room side, the closet door was slightly open, and it was unlatched. This had happened many times before and I decided to see what was going on. It was around 11:30 pm when I went into the living room with all the lights out except the glow of the light over the kitchen stove. I sat in my antique rocker across from the closet door, and after only a minute or two, it began to open slowly. A little at first then a little more, and a little more. Then it closed as though my presence startled them. It was not the captain; however, it was much smaller I could feel a little boy's presence. A year later, I was getting married and moving to Rockport, which is about a five-to-ten-minute drive from Gloucester. Let us head to Rockport where the Spirits are equally active, if not more so than in Gloucester. However, no matter where I live in New England there will always be a special connection between me, and the captain, and his house.

Five

Detour…

Let us take a short detour and go back to SeaSide Cemetery, and Flat Rocks. If you read my first book, (*Haunts of New England "They're Here, And Watching"*) you will remember our walk-through SeaSide Cemetery, onto the Path toward Flat Rocks, and the "Dark Shadow Men" along the way. Here are some more interesting pictures from that walk that I discovered.

The face within this Orb on the next page looks like "A Caveman" he is definitely pre-historic, or perhaps American Indian. The right side of his face is showing the eye, the nose and the shape of his mouth, draws me to my conclusions. You can see how others are watching us.

Look right above in the middle, there is a glowing half ring, inside the ring (Orb) is the face. You be the judge what does it look like to you?

I know of only three Paths leading from SeaSide Cemetery to Flat Rocks. I have been on each one of them, but am sure there are others. The first book are pictures taken on the most popular Path used, I refer to it as the 1st Path. The 2nd Path gave me a mystical feeling. Like in a blink of an eye, one could actually transport to some other Dimension. Never to be seen or heard from again. It is on this 2nd Path that the "Spirit Head" was watching the broken Television set.

Six

1st Path

As you can see, this Orb shines through even in the daylight. This is the exact spot where I took “The Spirit of Flat Rocks” picture. Now considering he may be an actual Gnome; here is where one can be found.

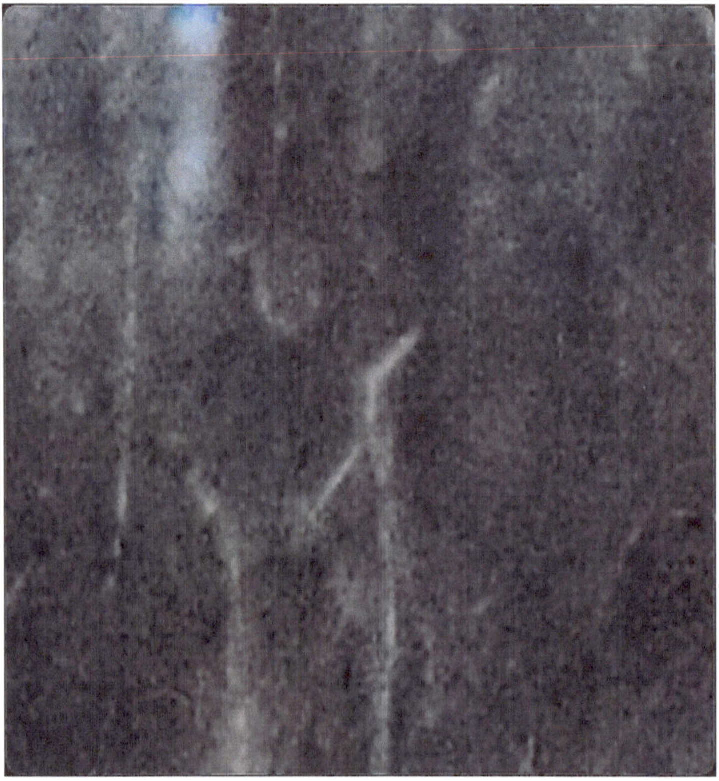

The Spirit of Flat Rocks

See the small Orange Orb above my head. There is also a larger white one behind me; there is a very mystical feeling here.

Flat Rocks

1st Path

Night falls on the 1st Path

See the abundance of Orbs and Ectoplasm in this shot. It can be very disturbing when you see this in your lens in the middle of the woods in the dark. This is SeaSide Cemetery at night!

One says that a Cemetery is the "resting place" for the body that grows tired, sick and old, yet it is certainly not the "final resting place" for our Soul as

I have learned and experienced, they are very active.

Some curious Spirits check us out as we check out our cameras.

1st Path toward Flat Rocks

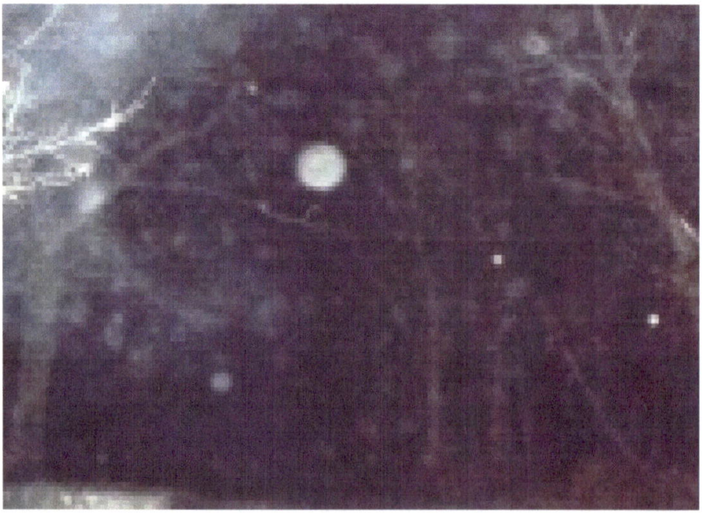

See the abundance of Orbs along this Path.

These two pictures were taken seconds apart, yet the first one, as you can see has "Ectoplasm". In an instant, it disappears, shown in the picture below it.

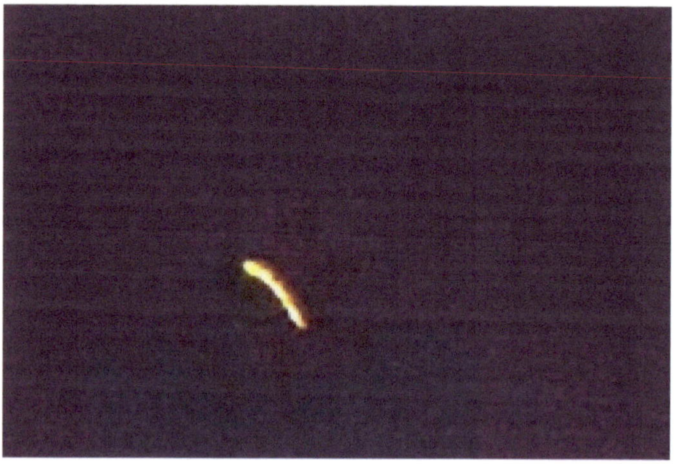

Here are my first pictures of “Energy Lights”,

caught in SeaSide Cemetery. Notice the strange

shapes and colors. How we cannot see this with the
naked eye is mind blowing to me.

Some curious Spirits escorted me back to my car.

Below are two pictures from my first book the two Orbs look to be the same as the one with the Demon on the hood of the car below.

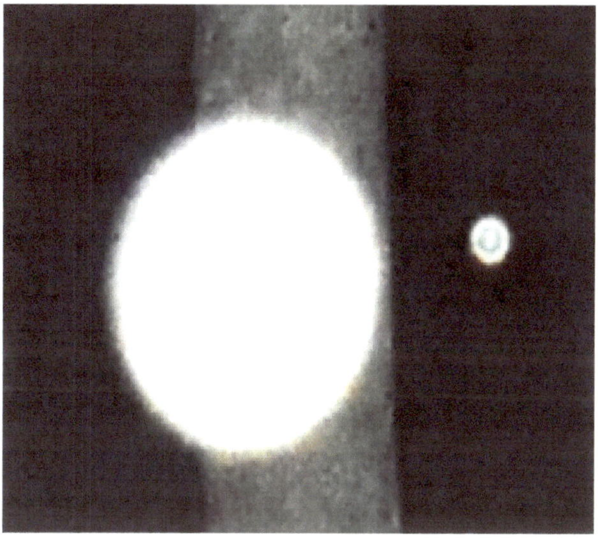

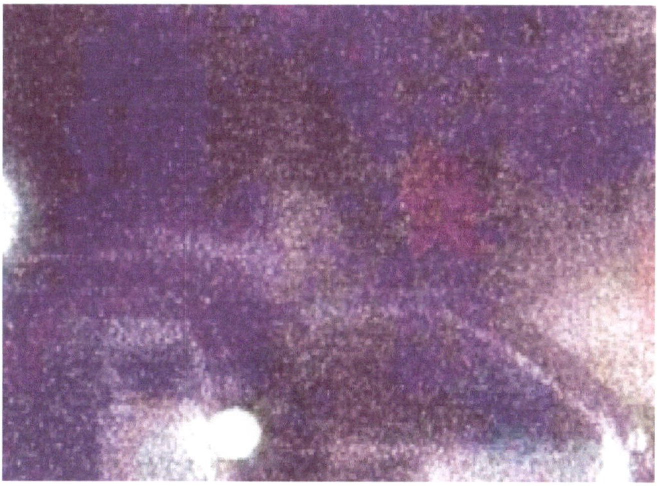

Seven

2nd Path

Springtime arrives and we follow the 2nd Path towards Flat Rocks…

Joe walks ahead to clear the way on the 2nd Path.

Spirit Head watches the Television set.

This picture really does blow my mind and deserves much more attention than I can give it. Who is this man? What is he doing, what is he observing? I wonder what era he lived. Why is he watching the Television? Then I see all the others around him looking back at me. How does this affect our World? These questions and others need answering.

This is that broken Television set which seems to work just fine in the Spirit World. What sort of message is this? This must make you wonder, “What the hell is going on here?” This tells me someone is trying to communicate something of great importance to us from these other Dimensions. There is great power and knowledge reaching out to us farther than our minds can imagine, and I believe these are not just random pictures of Ghost or

Spirits. I believe we need to find out what these Spirits are trying to say or trying to show us.

Daylight on this 2nd Path and the Spirits are active mingling in our World, watching us all the time.

People do not venture this path often. The pricker bushes grow out of control here; one can easily become tangled up if they are not careful.

The entrance to a Cave off the 2nd Path, which I did not go into, I do not like cramped tight spaces. Perhaps next summer I will try to brave it and take some video footage of anyone who lives inside.

Eight

Flat Rocks Revisited

Christopher (*my son*) joins me on one of my walks

and enjoys the beauty of Flat Rocks.

Treasures found

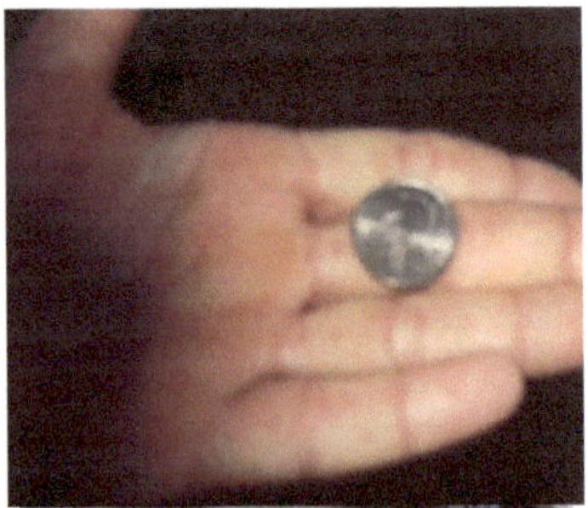

A Coin from long ago we could not see a date, and some type of Stone.

This Spoon makes me wonder who used it, and how and when it came to be in between the Rocks here. They may not seem like Treasures to you, but for me its pieces of New England History.

Nine

3rd Path

A few weeks later we decided to take the 3rd Path, which has not been trimmed in a while, a very long while, if ever. The branches have grown out of control here, and in order to walk we had to cut the bramble to clear a path. Something powerful lives here and it did not want us to intrude, as soon as we stepped into the brush, I got a headache and Joe's car alarm went off. Trying to turn off the car alarm from where we were was impossible, so Joe took off toward the car. I stayed behind, took two Tylenol, and kept taking pictures but the pictures did not come out properly. Only standing at the entrance of this 3rd Path (*shown below*), were they clear, once I stepped inside the Path it was over for the camera.

When Joe returned, we decided to split up. I went off to the left and Joe went deeper into the right, but we did not get far, the car alarm went off again. Joe went back a second time to turn off the alarm leaving me behind, this time alone deep within the 3rd Path. I began to feel the presence of multiple Spirits coming at me in all directions and I did not like it one bit. I tried to keep my cool and not panic, but then the wind began to pick up, slinging branches all around my face. I tried walking back to find the entrance, but it was hard to see, and the bramble and branches were on the attack. They were sticking me getting in my hair, tripping me up, anything to get me frustrated and scared. I am not ashamed to say, it was working, and I wanted off this 3rd Path, and where was Joe? It seemed like he had gone forever. Then I heard him in the distance, he was getting closer now and we both knew it was time to get out of there. I believe this Path to be the Heart of what lives in these Woods. These are what the pictures look like on the 3rd Path

This one almost looks "Alien" what do you think?

The top picture is very interesting indeed, there are a multitude of Spirits, but the bodies are small, maybe children. In the bottom right-hand corner and throughout this picture you will see faces and bodies, just keep looking, they will come into view.

Here is another picture on this 3rd Path its clear the Spirits here do not like their pictures taken. I am standing perfectly still I even leaned on a tree to make sure there was no movement and waited until there was no wind. Nevertheless, each time the pictures look distorted.

See what you can see…

We make it off the 3rd Path, I am more than ready to go home, and as usual, I get an escort back to my car.

Detour over, and we head to Rockport, where the Spirits are just as active if not more so than in Gloucester…

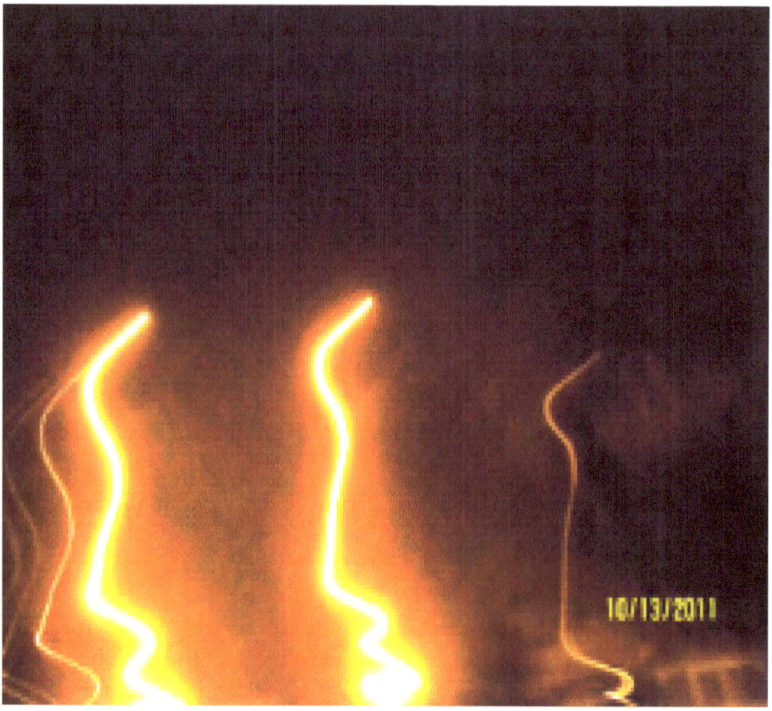

More strangely shaped Energy Lights in Rockport, Massachusetts

Ten

Wedding Day Heart

Moments before Joe and I said our vows, this first picture taken by his brother at City Hall in Gloucester, shows a strange dot at the base of my knees.

Here is a close up of my knees, This Orb or Orbs has formed a perfect Heart, truly amazing…

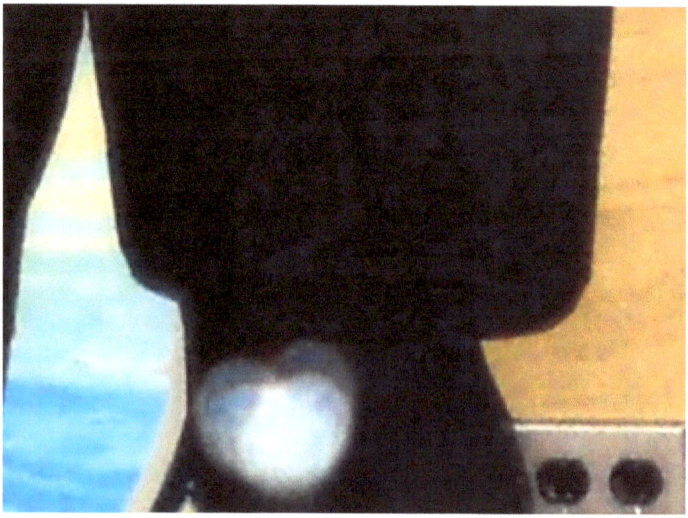

Eleven

Shadow Man Pays a Visit

Never let anyone tell you that Spirits will not follow you home, because I have positive proof that they do. Someone from one of the Paths from Seaside Cemetery came home with us, and I could feel the positive energy quickly drained from my home and my relationship with Joe. Shortly after our first visit to SeaSide Cemetery and Flat Rocks, we began to argue, a little at first, then more and more. There was a foul odor in the air at times, and I noticed that my cats were acting strange. At times, I would notice my cats looking at this spirit, as if someone were talking to them. Then they would lower their heads as though an invisible hand was petting them. Then other things began to occur. I would be alone in one room and would hear Joe ask me a question, as if he were right there with me. When I answered him and waited for a reply, or turned my head and expected to see him, he was never there. I would call out to him, and he would always be somewhere else either watching television or not in the house at all. Then "It" began to follow me around the house, I could tell by the way my cats looked at me when I entered a room. They even stopped following me around and sometimes would even hiss when I came near them. I purposely ignored this Spirit, hoping that it would just go away, but he was staying. A few nights before I decided to cleanse my house, I'd gone outside of my apartment in Rockport to take a few pictures, (*I will get the urge to do so, and I just go with it.*) This would be my first time taking pictures outside of this apartment building and I was curious to see what was there. After the very first picture, I snapped there was more activity than I expected, there were Orbs everywhere. Once again, I snapped and saw two eyes staring back at me, it was a coyote! I stumbled backwards, and kept snapping pictures as I made my way to the front door hoping the flash would keep it away, I was safe inside. I suddenly felt tired I took out the laptop to download the pictures. Afterwards I relaxed on the sofa with DeeDee, (*one of my cats*), and watched television. Then I felt something watching me from the hallway, and there was that foul smell again. DeeDee saw it also, she was looking at something in the hall, it was time to see what it was. I raised up slightly, got the camera from the table in front of me, and snapped a picture. When I saw the image in my lens, I jumped off the couch and started turning on lights, running down the hall toward the bedroom where Joe was. There was a man standing there in the hallway "Watching" me. I knew he came from SeaSide Cemetery.

Look to the left of the picture there is the outline of a man. There are others if you take a closer look. There is no worse feeling than that of being uncomfortable in your own home. After seeing that picture, I began to keep lights on throughout the house in the evenings. Walking down the hall I could feel him watching me, sometimes I could see his dark mist ahead of me, I had had enough of him, and it was time for him to go! I told Joe we needed to stay out of SeaSide Cemetery and Flat Rocks for a while, at least until I got rid of this Spirit. After cleansing my apartment and getting rid of that foul smell, life resumed, and things felt better.

Sometime later, I felt eyes on me again in that hallway, so I picked up my camera and took another picture and sure enough, something is still existing in that hallway in this Rockport apartment.

What appears to be white spots are faces emerging from the back room.

Twelve

Squiggly Creature

This next picture I took one summer evening in Rockport I got that urge to get outside now that the sun had set to take some pictures. I decided to take out the trash as well, so I grabbed the trash bag and my camera. I walked toward the dumpster and placed my bag inside, when suddenly, I got that strange feeling something was behind the dumpster. Not on the ground but it was high in the trees, so I took a picture of the trees behind the dumpster.

Look at this little animal, if anyone has ever seen a creature like this and know what it is, please contact me.

Thirteen

Big-Headed Rider

Another beautiful summer day in Gloucester, and after spending a day on the water, Joe and I headed home to our apartment in Rockport. I had to go outside and take some pictures now that the sun had set because there was something in the air. I saw so many Orbs and Energy Lights in my lens that it was no surprise when I saw these next pictures. There in the backseat of Joe's Ford Explorer sat a man with an unusually large head. Right in front of him, is another smaller man with dark hair sitting on his shoulder, looking right at me. The next day I called my oldest sister, Valerie who lives in Matteson, Illinois and told her about the Big-Headed Man, before I could finish my story, she gasped, and I knew she had something good to tell me. This is what she said, (*"The other day I was watching one of those Real Ghost Story Shows ((one of the many shows where every day people can share their own true accounts with the paranormal, as well as Celebrity Ghost Stories, where celebrities share unique tales of their own)) and there were these two men who collect vintage cars. They had about fifty cars in a garage. Every time they arrive in the morning things would be moved from the night before. Sometimes something would be taken off the walls and thrown to the floor. In the evenings on a few occasions, they could hear strange noises, so they decided to put up a camera to see what was going on, and in one of the vintage cars was a man with an unusually large head"*). I e-mailed her a copy of my picture, and it was in fact the same Big-

Headed Man, "Wow" I couldn't believe it really, my mind began to whirl, and my curiosity again was peaked, "Who is or was this Big-Headed Man, and why does he appear in automobiles?"

Notice the two images in the rear of the truck, there is the profile of a man with the unusually big head, and right under his ear on his shoulder, is the little man with dark hair looking right at me. Then behind the truck is some bulldozer type of equipment, but notice the white glow in the upper left side of the picture? Look a little closer it's a face, as to what type of face I am not sure. If you can, magnify this image, and any of my other images, you will see a multitude of faces…

Fourteen

House 13

Sometimes when you least expect it, something happens that makes you stop and go "WHOA!" On the very same night, that I captured the "Big Headed Man", across the street being built was one of those prefabricated homes. They pour a foundation then a shell of a house is placed onto the foundation, leaving the inside of the house empty. I stepped into the middle of the street and took a few pictures of the empty house, (*earlier that morning the shell of a house was set into the foundation*). When I looked in my lens, I saw eyes looking out from the windows inside the empty house.

After seeing these eyes in the window, I realized the white face in the bottom right windowpane is the same as the one above the bulldozer. Then looking in the bottom left windowpane, there are two people staring back at me, not to mention the people in both top windowpanes. "*House 13*" does not disappoint if comes complete with its own tenants.

Fifteen

The Peepers

The story I am about to share with you is a piece of history that I am sure we all would like to forget. Off Rt.133 in Gloucester there is this small bridge hidden in the woods. Once I crossed that bridge, the pain that slavery caused so many unfortunate soles to endure, lingers in the air. I captured a picture of some of these Slaves who seem to be afraid still, and I am a loss for words. When I first saw them in my lens, I immediately began to cry, I could now see who was trying so desperately to get my attention the first time I visited this house, over a year ago. A deep sadness engulfed me, I felt like I had abandoned them, but I was back, and they did get my attention! I met a woman at Lanes Cove, a beach in Lanesville in Gloucester, who had just moved into a house across that bridge along that stretch of road. During our conversation, she began to tell me how she felt multiple presences around her most of the time, especially from this small closet upstairs in her bedroom. When she first moved into the house, she and her son were sitting in the backyard, when he turned and her asked, "Mom, who is that man coming out of the house?" she turned toward the house and not seeing anyone, asked him what man he was talking about. He pointed and said, "The big black guy right there"! She watched as his finger trailed the Spirit come out of her house, into the yard next door toward the well and disappear. This finally made her ask a few questions in the neighborhood, and someone told her that her house, like many on this road, were once slaughterhouses. There was a tale of a slave that fit her son's description of the tall man, who drowned in this well either as he tried to hide from his master, or he was murdered, and his remains thrown down the well. She was also told that 3 feet of cement was poured in these cellars to cover the shackles, where slaves were chained while the Masters dined, bathed and ate above. (*And God only knows what else is buried in these damp cellars along this road*). I agreed to come over and take some pictures. I was of course extremely curious to see this place, but I had no idea what I was in for. It was a sunny autumn afternoon when I first visited her home, and as I turned off Rt. 133 onto this dirt road, my heart began to beat faster. When I turned my car toward the bridge, my stomach fell to my feet, and I felt sick. This road has so much dark energy you could cut it with a knife, my first impulse was to turn and go back, but I drove on. I pulled my car over to the side of my friend's house and broke out my camera, I could feel the energy all around. I went inside and took a few pictures, there were Orbs everywhere, and then my head began to ache. The more I snapped the camera the worse my head ached, I took two migraine pills, but it was too late, I had to leave. The air around me was so thick, it became hard to breathe, and then the light burned my eyes, my vision was reduced little to none. (*I suffer from migraines; I am light and energy sensitive, which means I get headaches from direct sunlight or bright lights, and when there are multiple Spirits trying to get my attention).* When I turned off the bridge and headed back out of the woods toward the highway, my headache ceased.

My vision began to clear. I knew there was something there desperately trying to make contact. Months went by and seasons changed, and about a year or so later, I found myself driving back across that bridge toward the house in the woods. Just like before as I crossed the bridge, my insides fell, only this time I came prepared. I had taken my migraine medicine before I left home, and I was sure I would not suffer another episode like last time. After visiting for a while, I got out my camera and went through the house taking pictures. The Spirits were very curious as the house was being remodeled. After walking around downstairs of the house, I decided to go up and check out that small closet. However, this was no closet. It was a small room, where I imagined slaves were being kept for certain pleasures. I captured Orbs everywhere and then my head began to ache. I immediately took another one of my migraine pills and continued taking pictures. I began to walk back downstairs and walked through the small hall back into the living room. Just then I noticed there was this mirror hanging on the wall by the front entrance, and an unseen force pulled me in its direction. I stood back a few feet, pointed my camera toward the mirror, and snapped a picture. In my lens, I saw something so special, yet mind blowing and sad at all at the same time. My eyes were watering, and it was hard for me to comprehend what I was seeing. I saw these slaves crunched down, right on the other side of the wall "Peeping" at me through the mirror. I was so close to these slaves, I am guessing 17th or 18th Century. They are watching me in the mirror along with a tall dark figure standing behind the slave in the orange hat. He may be their capturer in this dimension, or he may not want to show his face to me.

Many slaves received harsh treatments. They were starved, and beaten and treated like animals and their energy is still very strong along this stretch of road. I believe that if they remove the shackles from these cellars and perhaps dig up the bones of the dead, these Spirits could finally be at peace.

This is one the most “AMAZING” Picture of Spirits I have ever seen there is a large Orb in the left corner of the mirror, and the faces on the right are truly “ASTOUNDING”. As I said before, slave’s endured awful treatment and unfortunately, the stain on this stretch of road is so great it still lingers in the air. My thoughts…

My thoughts…

As I wrote these stories, my mind began to venture in directions on a much higher plane than before; I believe I am tapping into other Dimensions here on Cape Ann. The Spirits that I am capturing are active and existing in another realm. They obviously can see us, and soon I believe we will be able to see them with the naked eye. I believe the lining is getting thinner with each passing day. I watched a television program one evening, hosted by Mr. Morgan Freeman who spoke about "Wormholes". He spoke of "wormlike lights of energy". In many of my pictures, I have captured wormlike energy lights of all colors, different sizes and unbelievable shapes. I knew it was some sort of energy I now believe is in fact the same lights I saw on his program. Some of these energy lights have shadows of people in and under their glow, and the incredible speed at which they move can only be caught for split seconds and then they are gone. One question that comes to mind is "What Dimensions are being captured? It is more than one of that, I am sure. Some of the faces are smiling at me, they seem happy when, and where they are. Some are trying to hide, which makes me feel they think I can see them, because of my light. Others seem unaware of my presence, or they just do not care. Then there are the different feelings I get when I am in the presence of these "Spirits".

My next book They're Here, And Watching III – "DogTown Commons, and Beyond", takes you into DogTown where I experienced horrible feelings of dread. I began to see what I was feeling, and I did not want to see it. Moreover, immediately my head began to ache, I needed to leave, but not before a woman appeared. I knew there was something strange about the woman I knew she had died in these woods.

Since I began writing this particular book, I am constantly hearing people say to me they are not afraid of dying. Here is "positive proof" there is another existence in which we will be plunged once the body has expired. However, where we will be plunged, and what may be there to stop our journey home, we will not know until that day arrives. Some of the Spirit's, I have captured exude pure evil, and I would not want to end up along their paths. Then I wonder about these Energy Lights? Do the different colors and shapes put you on certain Spiritual levels? So many questions, will we ever know the answers while we exist in this day and time?

I have always felt that life on Earth is a Big School, and we all are being Watched, Tested, Judged, and get chances to get it right over, and over again. I believe this now more than ever. How well you lived your life determines where your Soul will end up. We do not know where it goes or for how long, so being afraid of the unknown is as common as air. Since I have been taking my pictures, some of the images I have caught, makes me want to live a full healthy happy God-fearing life. Treating others, the way I want to be treated. To be kind to strangers, and animals, being sensitive to my surroundings, enjoying, and being grateful for each day. I pray you do as well.

July 19, 2012

A book on local haunts By Gail McCarthy, Staff Writer

Gloucester Daily Times

---- When Antrella Grice first visited Cape Ann in 1989, she was entranced with the surroundings. When she moved here from Chicago in 2005, she became enthralled with the abundance of paranormal activity she discovered. These experiences are now documented in a self-published book, “They’re Here, And Watching.” Upon settling here, she began exploring Gloucester’s haunted tales and found two places where she said the Spirits are not camera shy. She plans to continue writing and incorporating her plentiful photographs into future books. Grice has studied communications and acting.

(Special Thanks to Gail McCarthy of The Gloucester Times.)

You can purchase my books through Amazon and Kindle books search for Antrella Grice.

Check me out for live stories at tiktok.com@antrella8

Check out some of my favorite pictures and videos on my Facebook page at [They're, Here and Watching @ Facebook.com]

Until next time, *Enjoy Life, Live Well, Do Unto Others As You'd Have Done To You Because,*

They're Here and Watching!!!

More about the Author

I am originally from Chicago, Illinois but now I reside in Gloucester, Massachusetts. I have studied Acting, Singing Journalism and Communications. I appeared in two movies, one with Molly Ringwald and another where I had the privilege to meet and sit alongside Mr. Sidney Poitier. I received a

speaking role in a Bud Lite Spotlight commercial and sang backup with Rick Howard and the Chicago Backstreets in a music video. I found my true passion, however, is my writing and exploring this World of ours and its many mysteries and shadows.

www.ingramcontent.com/pod-product-compliance
Ingram Content Group UK Ltd.
Pitfield, Milton Keynes, MK11 3LW, UK
UKHW061023310726
14090UKWH00023B/33
9781475273649